DELAWARE COMPOSITES DESIGN ENCYCLOPEDIA

INDEX TO
VOLUMES 1–6

Compiled by HARRY E. PEBLY

Reviewing Editors

LEIF A. CARLSSON, Ph.D.   JOHN W. GILLESPIE, JR., Ph.D.

CRC Press
Taylor & Francis Group
Boca Raton London New York

CRC Press is an imprint of the
Taylor & Francis Group, an **informa** business

Delaware Composites Design Encyclopedia—Index

First published in 1991 by Technomic Publishing Company, Inc.

Published 2019 by CRC Press
Taylor & Francis Group
6000 Broken Sound Parkway NW, Suite 300 Boca
Raton, FL 33487-2742

© 1991 by Taylor & Francis Group, LLC
CRC Press is an imprint of Taylor & Francis Group, an Informa business

First issued in paperback 2019

No claim to original U.S. Government works

ISBN-13: 978-0-367-45065-6 (pbk)
ISBN-13: 978-0-87762-705-0 (hbk)

**Visit the Taylor & Francis Web site at
http://www.taylorandfrancis.com**

**and the CRC Press Web site at
http://www.crcpress.com**

Library of Congress Card No. 89-51098

C O N T E N T S

CONTENTS

Volumes 1–6

VOLUME 1—MECHANICAL BEHAVIOR AND PROPERTIES OF COMPOSITE MATERIALS

Foreword   vii

1.1   INTRODUCTION/C. Zweben...3

    1.1.1   General Characteristics of Composite Materials   *3*

    1.1.2   Fibers   *4*

    1.1.3   Resins   *9*

    1.1.4   Fillers   *11*

    1.1.5   References   *13*

1.2   MECHANICS OF COMPOSITE MATERIALS/C. Zweben...............................17

    1.2.1   Introduction   *17*

    1.2.2   Unidirectional Composites   *17*

    1.2.3   Fabric Composites   *31*

    1.2.4   Discontinuous Fiber Composites   *37*

    1.2.5   References   *44*

1.3   STATIC STRENGTH AND ELASTIC PROPERTIES/C. Zweben...........................49

    1.3.1   Introduction   *49*

    1.3.2   An Overview of Material Properties   *49*

    1.3.3   Unidirectional Composites   *62*

    1.3.4   References   *69*

1.4   FATIGUE OF COMPOSITES/H. T. Hahn...73

    1.4.1   Introduction   *74*

    1.4.2   Unidirectional Continuous Fiber Composites   *75*

    1.4.3   Multidirectional Laminates   *85*

    1.4.4   Discontinuous Fiber Composites   *99*

    1.4.5   Life Prediction and Data Analysis   *107*

    1.4.6   Appendix: Weibull Distribution   *122*

    1.4.7   References   *122*

1.5   MECHANICS OF TWO-DIMENSIONAL WOVEN FABRIC COMPOSITES/T.-W. Chou........131

    1.5.1   Introduction   *131*

    1.5.2   Geometric Characteristics   *131*

    1.5.3   Methodology of Analysis   *132*

    1.5.4   The Mosaic Model   *133*

VOLUME 2—MICROMECHANICAL MATERIALS MODELING

Foreword   vii

2.1   INTRODUCTION TO ANISOTROPIC ELASTICITY/R. L. McCullough.......................3

   2.1.1   Elementary Material Descriptors   *3*
   2.1.2   Description of Material Behavior   *5*
   2.1.3   Directional Dependence   *10*
   2.1.4   Materials Symmetry   *18*
   2.1.5   Engineering Constants   *26*
   2.1.6   References   *35*

**2.2   SPECIAL APPLICATIONS OF MECHANICS OF
        ANISOTROPIC MATERIALS**/J. M. Whitney...39

   2.2.1   Cylindrical Orthotropy   *39*
   2.2.2   Plane Stress   *40*
   2.2.3   Shear Coupling Phenomena   *42*
   2.2.4   Expansional Strains   *43*
   2.2.5   Equations of Elasticity   *44*
   2.2.6   References   *45*

**2.3   MICRO-MODELS FOR COMPOSITE MATERIALS—
        CONTINUOUS FIBER COMPOSITES**/R. L. McCullough..............................49

   2.3.1   Micromechanical Models of Unidirectional Continuous Fiber Composites   *50*
   2.3.2   Mechanics of Materials Analysis of In-Plane Engineering Constants   *56*
   2.3.3   Self-Consistent Field Models   *65*
   2.3.4   Bounding Methods   *73*
   2.3.5   Semi-Empirical Models—The Halpin-Tsai Relationship   *82*
   2.3.6   Summary and Recommendations   *86*
   2.3.7   References   *89*

**2.4   MICRO-MODELS FOR COMPOSITE MATERIALS—PARTICULATE AND
        DISCONTINUOUS FIBER COMPOSITES**/R. L. McCullough...............................93

   2.4.1   Review of Models for Particulate Filled Systems   *99*
   2.4.2   Improved Combining Rules for Particulate Composites   *108*
   2.4.3   Micro-Models for Sheet Molding Materials   *116*
   2.4.4   References   *141*

2.5 **INTRODUCTION TO VISCOELASTICITY**/R. L. McCullough **145**

    2.5.1 Characteristics of Viscoelastic Behavior *146*

    2.5.2 Mechanical Analogs *157*

    2.5.3 Mathematical Models *166*

    2.5.4 Summary of Linear Viscoelasticity *182*

    2.5.5 Viscoelasticity Annotated Bibliography *186*

    2.5.6 References *186*

2.6 **MICRO-MODELS FOR PREDICTING VISCOELASTIC BEHAVIOR**/R. L. McCullough **189**

    2.6.1 Review of Approaches to Predicting Viscoelastic Behavior of Composite Materials *189*

    2.6.2 Fundamental Formulation *190*

    2.6.3 Generalized Viscoelastic Models *199*

    2.6.4 References *200*

2.7 **TRANSPORT PROPERTIES**/R. L. McCullough **205**

    2.7.1 Elementary Material Descriptors *206*

    2.7.2 Generalized Transport Behavior *215*

    2.7.3 Review of Combining Rules for Transport Properties *218*

    2.7.4 Effective Chain Model for Transport Properties *222*

    2.7.5 Comparison of Model Predictions with Characterization Data *225*

    2.7.6 References *230*

VOLUME 3—PROCESSING AND FABRICATION TECHNOLOGY

Foreword   ix

3.1   **HISTORICAL BACKGROUND**/M. G. Bader..3

3.2   **REINFORCEMENTS**/M. G. Bader...7
    3.2.1   Glass Fibers   *7*
    3.2.2   Graphite/Carbon Fibers   *8*
    3.2.3   Synthetic Organic Fibers   *8*
    3.2.4   The Form of the Reinforcement and Its Preparation for Molding   *8*
    3.2.5   References   *11*

3.3   **RESIN SYSTEMS**/W. Smith...15
    3.3.1   Introduction   *15*
    3.3.2   Polyester Resins   *15*
    3.3.3   Epoxy Resins   *27*
    3.3.4   Thermoplastic Matrix Resins   *39*
    3.3.5   Polyimides   *61*
    3.3.6   Phenolic Resins   *78*
    3.3.7   References   *84*

3.4   **MOLDING PROCESSES—AN OVERVIEW**/M. G. Bader.................................89
    3.4.1   Summary of Basic Types   *89*
    3.4.2   Summary of Economic Factors for Basic Systems   *97*
    3.4.3   References   *101*

3.5   **OPEN MOLD LAMINATIONS—CONTACT MOLDING**/M. G. Bader.....................105
    3.5.1   Introduction   *105*
    3.5.2   The Mold   *105*
    3.5.3   Choice of Reinforcement   *106*
    3.5.4   Choice of Resin   *109*
    3.5.5   Hand Lay-up   *112*
    3.5.6   Spray-up Processes   *114*
    3.5.7   References   *115*

3.6   **COMPRESSION AND TRANSFER MOLDING OF REINFORCED THERMOSET RESINS**/M. G. Bader...119
    3.6.1   Classical Thermoset Resins and Compounds   *119*
    3.6.2   Thermosetting Resin Systems   *121*

3.6.3   Fillers and Reinforcements for Thermoset Systems   *122*

3.6.4   Compression Molding   *124*

3.6.5   Transfer Molding   *128*

3.6.6   References   *129*

3.7   **SMC AND BMC PRESS MOLDING TECHNOLOGY**/A. B. Isham **133**

3.7.1   Introduction   *133*

3.7.2   Selection of Materials   *133*

3.7.3   Compounding Technology   *134*

3.7.4   Molding Technology   *139*

3.7.5   Mechanical Properties   *143*

3.7.6   References   *145*

3.8   **INJECTION MOLDING**/M. G. Bader **149**

3.8.1   General Principles   *149*

3.8.2   The Principles of Operation of the Injection Molding Machine   *149*

3.8.3   Design and Control of Injection Molding Machines   *152*

3.8.4   Machine Operating Settings   *155*

3.8.5   General Aspects of Mold Design   *158*

3.8.6   Special Considerations for Molding Filled Materials   *158*

3.8.7   Fiber Reinforced Thermoplastics (FRTP) Properties   *159*

3.8.8   Compounding FRTP   *161*

3.8.9   Mold Filling and Flow-Induced Fiber Orientation   *163*

3.8.10   Precision Molding   *165*

3.8.11   Fiber Reinforced Thermoplastics (FRTP) Summary   *166*

3.8.12   References   *169*

3.8.13   Bibliography   *169*

3.9   **INJECTION MOLDING FIBER REINFORCED (F.R.)**
        THERMOSETS AND BMC/M. G. Bader **173**

3.9.1   Introduction   *173*

3.9.2   Fundamental Principles   *173*

3.9.3   Thermosetting Compounds for Injection Molding   *173*

3.9.4   Injection Molding Machines for Thermosets   *174*

3.9.5   The Injection Molding Operation   *174*

3.9.6   Mold and Cavity Design   *175*

3.9.7   Summary—Injection Molding of Conventional Fiber Reinforced Thermosets   *175*

3.9.8   Injection Molding of Bulk Molding Compounds (BMCs)   *176*

3.9.9   Molds for BMC Injection Molding   *176*

3.9.10   Injection-Compression Molding   *177*

3.9.11   Summary and Comparison of Methods for Molding Fiber Reinforced
         Thermosets and BMC   *177*

3.9.12   Reference   *178*

VOLUME 4—FAILURE ANALYSIS OF COMPOSITE MATERIALS

Foreword   vii

4.1   **INTRODUCTION**/P. W. R. Beaumont, J. M. Schultz..3
    4.1.1   Reference   *6*

4.2   **FAILURE DETERMINED BY BULK PROPERTIES**/P. W. R. Beaumont, J. M. Schultz...........9
    4.2.1   Single and Multiple Cracking   *9*
    4.2.2   Stress-Strain Relationships   *12*
    4.2.3   References   *20*

4.3   **MICROMECHANISMS OF FRACTURE IN CONTINUOUS FIBER COMPOSITES—PART I: THEORY**/P. W. R. Beaumont, J. M. Schultz........................23
    4.3.1   Fracture Stress and Energy   *24*
    4.3.2   Comparison between Models of Fracture Mechanisms   *35*
    4.3.3   References   *36*

4.4   **MICROMECHANISMS OF FRACTURE IN CONTINUOUS FIBER COMPOSITES—PART II: DATA**/P. W. R. Beaumont, J. M. Schultz.........................41
    4.4.1   Method of Analysis of Fiber Debond Length and Fiber Pull-out Length   *43*
    4.4.2   Development of Calculated Work of Fracture Diagrams   *50*
    4.4.3   Work of Fracture of Structural Fibrous Composites   *51*
    4.4.4   Summary and Implications   *52*
    4.4.5   References   *52*

4.5   **ENVIRONMENTAL FAILURE**/P. W. R. Beaumont, J. M. Schultz............................55
    4.5.1   General Characteristics of Temperature/Time Aging   *55*
    4.5.2   "Extreme Value" Statistical Analysis   *59*
    4.5.3   Kinetics of Moisture Absorption   *63*
    4.5.4   Aggressive Environments   *63*
    4.5.5   References   *65*

4.6   **STATISTICAL ASPECTS OF FRACTURE**/P. W. R. Beaumont, J. M. Schultz...................69
    4.6.1   Weibull Statistics   *70*
    4.6.2   Evaluation of Flaw Distribution Parameters   *71*
    4.6.3   Specimen Size Effects   *73*
    4.6.4   Failure Stresses Related to Manufacturing Processes and Test Methods   *74*
    4.6.5   Chain-of-Bundles Model   *77*
    4.6.6   References   *82*

4.7   **MACROSCOPIC FRACTURE THEORIES**/P. W. R. Beaumont, J. M. Schultz 85

4.7.1   Fracture Mechanics Approach to Crack Propagation   *85*

4.7.2   Fracture Behavior of Unidirectional Continuous Fiber Composites—LEFM   *97*

4.7.3   A Shear Lag Approach to Fracture Stress   *99*

4.7.4   Fracture Behavior Based on Damage Zones   *101*

4.7.5   Appendix A. Common Unit Conversions   *110*

4.7.6   Appendix B. The Westergaard Stress Function Viewpoint for Isotropic and Orthotropic Two-Dimensional Problems   *110*

4.7.7   References   *114*

4.8   **FRACTOGRAPHY**/P. W. R. Beaumont, J. M. Schultz 119

4.8.1   Continuous Carbon-Fiber Composites   *119*

4.8.2   Continuous Glass-Fiber Composites   *128*

4.8.3   Injection Molding Compounds   *130*

4.8.4   Sheet Molding Compounds   *135*

4.8.5   References   *135*

4.9   **FRICTION AND WEAR OF POLYMER COMPOSITES**/K. Friedrich 139

4.9.1   Introduction   *141*

4.9.2   Fundamental Aspects of Friction and Wear   *142*

4.9.3   Comparative Friction and Wear Properties of Composite Materials   *165*

4.9.4   Friction and Wear Data of Selected Polymer Composite Systems   *179*

4.9.5   Conclusions   *202*

4.9.6   References   *203*

VOLUME 5—DESIGN STUDIES

Foreword   vii

5.1 A TYPICAL COMPOSITES DESIGN/K. T. Kedward 3

5.2 BASIC CONFIGURATIONS/K. T. Kedward 9
   5.2.1 Straight Beams   *9*
   5.2.2 Curved Beams   *12*
   5.2.3 Vibratory Response   *14*
   5.2.4 Thin Plates   *18*
   5.2.5 Design of Joints   *21*
   5.2.6 References   *42*

**5.3 APPLICATION OF NUMERICAL METHODS TO
        COMPOSITE STRUCTURES**/K. T. Kedward 47
   5.3.1 ANSYS Engineering Analysis System   *47*
   5.3.2 BOSOR 4 Shell of Revolution Code   *50*
   5.3.3 MARC Linear & Nonlinear General Purpose Program   *51*
   5.3.4 NASTRAN (NASA Structural Analysis Program)   *54*
   5.3.5 References   *55*

5.4 LAMINATION THEORY AND FREE EDGE STRESS ANALYSIS/J. M. Whitney 59
   5.4.1 Basic Assumptions of Lamination Theory   *59*
   5.4.2 Laminate Constitutive Relations   *60*
   5.4.3 Effective Elastic Constants of a Laminate   *65*
   5.4.4 Stresses in a Laminate   *67*
   5.4.5 Expansional Strains in a Laminate (Moisture, Temperature)   *71*
   5.4.6 Transport Properties for a Laminate   *74*
   5.4.7 Laminate Analogy for Short-Fiber Composites   *75*
   5.4.8 Edge Effects in Composite Laminates   *77*
   5.4.9 Appendix 1. Bending Stiffnesses for Angle-Ply Plates   *93*
   5.4.10 Appendix 2. Coupling Stiffnesses for Angle-Ply Plates   *94*
   5.4.11 Appendix 3. Effective Laminate Diffusivities   *96*
   5.4.12 Appendix 4. Approximate Free Edge Stresses   *98*
   5.4.13 References   *101*

5.5   STRENGTH ANALYSIS/J. M. Whitney **105**

   5.5.1   Strength Criteria for Orthotropic Homogeneous Materials   *105*

   5.5.2   Strength Criteria for Laminated Orthotropic Materials   *114*

   5.5.3   References   *121*

5.6   ANISOTROPIC PLATE AND BEAM THEORY/J. M. Whitney **125**

   5.6.1   Governing Equations for Heterogeneous Anisotropic Plates   *125*

   5.6.2   Displacement Formulation of Governing Equations   *127*

   5.6.3   Stress Function and Transverse Displacement Formulation   *128*

   5.6.4   Appendix. Energy Principles Applied to Anisotropic Plates   *129*

   5.6.5   References   *136*

5.7   ONE-DIMENSIONAL THEORY OF ANISOTROPIC PLATES AND BEAMS/J. M. Whitney **139**

   5.7.1   Cylindrical Bending   *139*

   5.7.2   Anisotropic Beam Theory   *143*

   5.7.3   Appendix. Determination of k Factors for Shear Deformation Theory   *155*

   5.7.4   References   *156*

5.8   SOLUTIONS TO LAMINATED PLATE PROBLEMS/J. M. Whitney **159**

   5.8.1   Bending under Transverse Loading   *159*

   5.8.2   Buckling Analysis   *171*

   5.8.3   Free Vibration   *193*

   5.8.4   Appendix 1. The Galerkin Method   *201*

   5.8.5   Appendix 2. Fourier Series Solution of Boundary Value Problems   *202*

   5.8.6   References   *204*

5.9   EXAMPLE DESIGN STUDIES/K. T. Kedward **207**

   5.9.1   Truck Frame Rail   *207*

   5.9.2   Drive Shaft   *214*

   5.9.3   Leaf Spring   *219*

   5.9.4   Door Intrusion Beam   *226*

   5.9.5   Automotive Bumper   *234*

   5.9.6   References   *243*

VOLUME 6—TEST METHODS

Foreword   vii

6.1   TEST METHODS/R. B. Pipes..3

    6.1.1   Constituent Test Methods   *5*

    6.1.2   Physical Property Test Methods   *7*

    6.1.3   Environmental Testing   *11*

    6.1.4   Composite Test Methods   *19*

    6.1.5   References   *53*

**6.2   ULTRASONIC NONDESTRUCTIVE EVALUATION TECHNIQUES FOR
COMPOSITE MATERIALS**/R. A. Blake, Jr..57

    6.2.1   Ultrasonic Inspection System Description   *57*

    6.2.2   Basic Types of Scans   *58*

    6.2.3   Amplitude Signature Evaluation   *71*

    6.2.4   Delamination Characterization   *73*

    6.2.5   Transducer Considerations   *75*

    6.2.6   Digital Systems Approach   *79*

    6.2.7   Appendix: Selected Examples of Composite Material C-Scans   *109*

    6.2.8   References   *109*

**6.3   INTERLAMINAR FRACTURE OF LAMINATED
COMPOSITE MATERIALS**/J. W. Gillespie, Jr., L. A. Carlsson..............................113

    6.3.1   Introduction to Interlaminar Fracture Mechanics   *113*

    6.3.2   Mode I Interlaminar Fracture   *119*

    6.3.3   Mode II Interlaminar Fracture   *126*

    6.3.4   Mode III Interlaminar Fracture   *146*

    6.3.5   References   *158*

I N D E X

Volumes 1–6

A

A-basis, statistical allowable **6**:4–5

A-scan, ultrasonic NDE **6**:58–61

Abrasive wear. See also *Wear mechanism.*
   fiber reinforced thermoset **4**:191–201
   general conclusions **4**:203

Abrasiveness, properties, related **4**:174–179

Accelerated conditioning **6**:15–19

Additives. See also *Filler.*
   contact molding **3**:111

Adhesive bonding
   graphite/polysulfone **3**:60–61
   joints **5**:27–32
   thermoplastic composites **3**:60–61

Aging, strength degradation **1**:112–114; **4**:55–58
   temperature/time **4**:55–58

Alkyd resins **3**:122

Alumina/epoxy, representative properties **1**:69

Amino resins **3**:122

Amontons-Coulomb Laws **4**:143

Amplitude signature, NDE evaluation **6**:71

Analog system, ultrasonic inspection **6**:57

Analysis. See also *Design; Model.*
   beam theory **6**:131–136
   critical buckling load **5**:171–193
   fatigue data **1**:114–117
   Fourier, NDE, digital system **6**:81–91
   free vibration, Fourier series **5**:197–198
   free vibration, Galerkin method **5**:195–201
   free vibration, laminated plate **5**:193–201
   max strain criterion **5**:106
   max stress criterion **5**:105–106
   models, homogeneous materials **5**:105–114
   quadratic interaction criterion **5**:106–114
   shear lag, LEFM **4**:99–101

Analysis, statistical
   extreme value **4**:59–62
   failure data **4**:46

Analysis, structural
   ANSYS computer code **5**:47–50
   automotive bumper **5**:237–243
   BOSOR computer code **5**:50–51
   door intrusion beam **5**:228–234
   MARC computer code **5**:51–54
   NASTRAN computer code **5**:54–55
   truck drive shaft **5**:214–219
   truck frame rail **5**:211–214
   truck leaf spring **5**:221–226

Angle-ply plate
   bending stiffness **5**:93–94
   fatigue life prediction **1**:109–122

Anisotropic elasticity. See *Elasticity, anisotropic.*

Anisotropic material
   cylindrical orthotropy **2**:39–40
   defined **2**:3
   directional dependence **2**:10–18
   elementary material descriptors **2**:3–5
   engineering constants **2**:26–35
   expansional strains **2**:43–44
   Hooke's Law **2**:4–5, 7–11
   load deformation characteristics **2**:3, 5–10
   materials symmetry **2**:18–26
   plane stress **2**:40–42
   shear coupling **2**:42–43

Anisotropic plate
   bending and buckling **5**:125–129, 142–143
   cylindrical bending **5**:139–143
   displacement formulation **5**:127–129
   one-dimensional theory **5**:139–143
   stress function formulation **5**:128–129
   vibration **5**:125–129, 142–143

ANSYS computer code analysis **5**:47–50

Aramid/epoxy
   fatigue **1**:78–82
   representative properties **1**:67

Aramid fiber. See *Fiber; Kevlar; Kevlar aramid fiber.*

Asbestos
   fabric **1**:34
   fiber **1**:4
ASTM test methods, materials
      characterization **6**:4
Attachment
   angle bracket **5**:41–42
   pin loaded lug **5**:37–39
   strap **5**:39–41
Autoclave molding **3**:95–96
Automotive components, design
   bumper **5**:234–243
   door intrusion beam **5**:227–234
   drive shaft **5**:214–219
   frame rail **5**:207–214
   leaf spring **5**:219–226

B

B-basis, statistical allowable **6**:4–5
B-scan, ultrasonic NDE **6**:62
Beam. See also *Plate.*
   anisotropic, one dimensional analysis . . .**5**:143–156
   bending equations **5**:143–147
   curved, stress calculation **5**:12–14
   intrusion, door, design study **5**:226–234
   resonant frequencies **5**:14–16
   shear specimen **5**:148–149
   straight, shear failure **5**:9
   straight, stress calculation **5**:9–11
   torsional frequency **5**:16–18
Beam-to-joint transition **5**:19–21
Bearings, friction and wear **4**:165–167
Bending
   anisotropic plate **5**:125–129, 142–143
   beam, equations **5**:143–147
   cylindrical, anisotropic plate **5**:139–143
   four point **5**:151–154
   Fourier series **5**:162–166, 202–203
   Galerkin method **5**:163–166, 201–202
   plate, transverse loading **5**:159–171
   stress, quasi-isotropic laminate **5**:68–71
   three point **5**:145–154
Bending stiffness, angle-ply plate **5**:93–94
Bisphenol A/fumarate resin **3**:22–23
BMC. See *Bulk molding compound.*

Bolted joint. See *Joint, mechanically fastened.*
Boron/epoxy
   fatigue **1**:77–78
   representative properties **1**:68
   stress-strain behavior **6**:23
BOSOR, computer code analysis **5**:50–51
Bounding methods, material behavior **2**:73–82
Buckling, anisotropic plate **5**:125–129, 142–143
   critical, analysis **5**:171–193
   Fourier series **5**:181–187,   202–203
   Galerkin method **5**:178–181, 185–187, 201–202
   one-dimensional analysis **5**:154–155
   symmetric laminate **5**:178–187
   unsymmetric laminate **5**:187–193
Bulk molding compound (BMC)
   composition **3**:133–134
   compounding **3**:139
   injection molding **3**:176–178
   mechanical properties **3**:143–144
   processing **3**:93, 139–143, 176–177
   processing influence on properties **3**:145
Bumper, automotive, design study **5**:234–243

C

C-scan
   consolidation **6**:102–109
   microstructure **6**:102–109
   ultrasonic NDE **6**:58, 62–78, 93–109
Calcium carbonate filler **1**:12–13
Carbon. See *Graphite.*
Carbon-carbon composite **1**:34
Carbon-fiber composite. See also *Graphite.*
   crack propagation **4**:125–128
   epoxy matrix, failure data **4**:76–77
   flexural failure **4**:122–124
   fracture toughness data **4**:97–100
   friction and wear **4**:167, 170
   impact failure **4**:125
   morphology, failure surface **4**:119–128
   shear strength **4**:127–128
   tensile failure **4**:119, 122
Celanese compression test **6**:29–30, 32
Characterization
   composite constituents **6**:5–7
   given material system **6**:3

Chemical composition. See *Chemical structure.*
Chemical resistance
    polyester **3**:24
    polyetherether ketone **3**:48
    polyphenylene **3**:47
    polysulfone **3**:47
Chemical structure
    epoxy **3**:27–30, 40
    high temp. thermoplastic resins **3**:43–44
    phenolic **3**:79–82
    poly(amide-imide) **3**:41
    polycarbonate **3**:41
    polyester **3**:15–16
    polyetherether ketone **3**:41, 44
    polyphenylene sulfide **3**:41, 44
    polysulfone **3**:41
Cloth. See *Fabric.*
Coefficient of moisture expansion **6**:9–10
Coefficient of thermal expansion **6**:8–9
    bismaleimide **3**:77
    polyester resin **3**:25–26
Cohesive wear. See *Wear mechanism.*
Combining rules
    composite materials
        performance **2**:189, 199–200
    definition **2**:93
    particulate composites **2**:93, 108–116
    polyblend **2**:115
Commercial availability
    filament winding equipment **3**:201
    phenolic prepreg **3**:84
    polyester resin **3**:26–27
    polyimide resin **3**:78
Composite design. See also *Design.*
    methodology **5**:3–5
    overview **5**:3–5
    trade-off study **5**:3
Composite material(s). See also *Anisotropic*
    materials; Discontinuous fiber composite;
    Laminate; Material behavior; Particulate
    filled composite; Unidirectional composite;
    and specific combinations, e.g. *Graphite/epoxy.*
    anisotropy **1**:3–4
    compression tests **6**:29–35
    constituent characterization **6**:5–7
    definition, schematic **2**:95
    design allowables **1**:59–60
    elastic and static strength **1**:49–69

    flexure test **6**:21–27
    general characteristics **1**:3–4
    heterogeneity **1**:3
    mechanical behavior **1**:17–19
    property variation **1**:3; **2**:55–59
    shear tests **6**:35–48
    tensile test **6**:19–21
    tensile test, off-axis **6**:27–29
Composite performance, essential
    constituents **4**:90
Compression molding
    alkyd resins **3**:122
    amino resins **3**:122
    fillers and reinforcements **3**:122–124
    introduction **3**:93, 119
    machinery **3**:127, 140–141
    mold design **3**:126–127, 141
    molding cycle **3**:125–126, 142–143, 177
    phenolic resins **3**:119–121, 128
    post-molding **3**:128
    pre-molding operations **3**:125
    special resins **3**:122
Compression test, composite **6**:29–35
    Celanese test **6**:29–30, 32
    IITRI test **6**:29, 31–32
    Lockheed test **6**:32
    NIST test **6**:29
    Northrup test **6**:32
    sandwich **6**:32–35
    SWRI test **6**:32
Compressive strength
    unidirectional composite **1**:27–28
Computer code(s)
    analysis **5**:47–55
    ANSYS **5**:47–50
    BOSOR **5**:50–51
    MARC **5**:51–54
    NASTRAN **5**:54–55
Consolidation, C-scan **6**:102–109
Constituent materials. See *Fiber; Filament;*
    Resin; Roving.
Constitutive relationships. See *Combining rules.*
Contact molding
    additives and promoters **3**:111
    cure **3**:111–112
    gel coat **3**:114
    glass fiber **3**:106–108
    graphite fiber **3**:107, 109

Contact molding (*continued*)

    introduction **3**:89–90, 105

    Kevlar fiber **3**:107, 109

    polyester, unsaturated **3**:109–111

    tooling **3**:105–106

    vinyl ester **3**:112

Corrosion wear. See *Wear mechanism.*

Cost

    vs availability, fibers **3**:97, 100–101

    vs availability, resins **3**:97–98, 100–101

Coupling

    bending-stretching **5**:62–63

    bending-twisting **5**:63–64

Coupling stiffness, angle ply plate **5**:94–96

Crack loading modes **4**:85, 87

Crack propagation

    carbon fiber composite **4**:125–128

    fracture mechanics approach **4**:85–97

    rate measurement **4**:94–95

Crack rail shear test **6**:147–158

Cracking

    matrix **4**:33–34

    single and multiple **4**:9–12

    stress calculation **4**:33–34

Criterion

    Hoffman, interaction coefficient **5**:108–109

    maximum strain **5**:106

    maximum stress **5**:105–106

    quadratic interaction **5**:106–114

    Tsai-Hill, interaction coefficient **5**:108

Curing

    contact molding **3**:111–112

    epoxy resin **3**:30–36

    hand lay-up **3**:112–114

    phenolic resin **3**:82–83

    polyester resin **3**:19–22

    polyimides **3**:61–67

    pultrusion **3**:211

Curing agents

    epoxy **3**:30–33

    phenolics **3**:82–83

    polyester **3**:21–22, 111

    polyimides **3**:61–67

D

Damage model, fracture behavior **4**:101–110

DAP resin. See *Alkyd.*

Data analysis

    S-N characterization **1**:117–120

    strength degradation, fatigue **1**:114–117

Debonding, fiber matrix **4**:25–31, 58

Delamination, NDE **6**:72–75

Density, test method **6**:7

Design

    adhesively bonded joint **5**:27–31

    attachments **5**:36–42

    automotive bumper **5**:234–243

    composite methodology **5**:3–5

    door intrusion beam **5**:227–234

    mechanically fastened joint **5**:23–26, 31–42

    preliminary, representative property

        data **1**:49–69

    pultrusion structures **3**:212–213

    sheet molding compound **2**:137–141

    strap attachment **5**:39–41

    trade-off study **5**:3

    truck drive shaft **5**:214–219

    truck frame rail **5**:211–214

    truck leaf spring **5**:219–226

Design allowables

    A basis **1**:59–60, 69

    B basis **1**:59–60, 69

    S basis **1**:59

    compared to aluminum **1**:60

Design Guide, Advanced Composites,

    U.S. Air Force **1**:51

Dielectric constant **2**:208–211

Dielectric permittivity. See *Dielectric constant.*

Diffusion

    coefficient **6**:12–14

    one-dimensional **5**:96–97

    two-dimensional **5**:97–98

Diffusivity. See *Transport property.*

Digital system, NDE **6**:79–97

    Fourier analysis **6**:81–91

    image processing **6**:92–97

Discontinuous fiber composite. See also *Sheet*

    molding compound.

    factors affecting properties **1**:37–38

    fatigue **1**:99–107

    fiber length effect **1**:39–44

    random orientation **1**:43–44

Displacement formulation, anisotropic

    plate **5**:127–129

Dither technique, NDE digital system **6**:93–100

Double cantilever beam
   interlaminar fracture **6**: 119–126
   R-curve effect **6**: 125–126
Doweave **1**: 33
Drive shaft, light truck, design study **5**: 214–219

E

Economics. See also *Cost.*
   of molding processes **3**: 99–101
Edge effect
   interlaminar stress **5**: 77–93
Edge stress **5**: 98–101
Elastic constants. See also *Engineering constants.*
   laminates **5**: 65–66
   unidirectional composites **1**: 19–23
Elasticity, anisotropic. See also *Anisotropic*
   materials.
   cylindrical orthotropy **2**: 39–40
   equations of elasticity **2**: 44–45
   expansional strains **2**: 43–44
   Hooke's Law, generalized **2**: 4–5, 7–11
   introduction **2**: 3
   plane stress **2**: 40–42
   shear coupling **2**: 42–43
Electrical conductivity **2**: 207–211
Electrical diffusivity **2**: 213–214
Electrical properties
   poly(amide-imide) **3**: 51
   polycarbonate **3**: 51
   polyimide **3**: 76–78
   polyphenylene sulfide **3**: 51
   polysulfone **3**: 50–51
Electrical resistivity. See *Electrical conductivity.*
End notch flexure test
   beam theory analysis **6**: 131–136
   interlaminar fracture **6**: 127–146
   nonlinear fracture toughness **6**: 143–146
Energy principle
   application plate **5**: 129–136
Energy release rate **6**: 115–119
Engineering constants, composite
   materials **2**: 26–35, 56–65
   symmetry reductions **2**: 32–35
   transformations **2**: 28–32
Engineering constants, in-plane
   expansional strains **2**: 61, 68
   Poisson's Ratio **2**: 52, 59, 67

   shear modulus **2**: 52, 60, 67
   shear modulus, transverse **2**: 68
   transport properties **2**: 69–71
   unidirectional materials **2**: 56–72
   Young's modulus, longitudinal **2**: 56–58, 67
   Young's modulus, transverse **2**: 60
Environmental failure **4**: 55–65
Environmental resistance
   polyimides **3**: 75–77
Epoxy matrix composite. See also *specific fiber.*
   combinations **1**: 65–69
Epoxy resin
   composition **3**: 27–30, 40
   cured properties **3**: 35–39
   curing **3**: 30–36
   curing agents **3**: 30–33
   for compression molding **3**: 122
   moisture absorption **3**: 39
   vs polysulfone **3**: 42, 53–55
Evaluation. See *Inspection.*
Expansional strains **5**: 71–74
   anisotropic materials **2**: 43–44

F

Fabric composite(s). See also *Fabric, woven,*
   2-dimensional composite.
   general description **1**: 31–34
   model, thermo-mechanical
     behavior **1**: 133–140, 143–149
   reinforcement mechanics **1**: 34–37
   3-dimensional **1**: 34
Fabric, woven, 2-dimensional composite
   advantages **1**: 131
   analysis methodology **1**: 132–133
   bridging model **1**: 139–140, 143–149
   crimp (fiber undulation)
     model **1**: 136–139, 143–149
   geometric patterns **1**: 131–132
   knee behavior analysis **1**: 140–143
   stress-strain behavior **1**: 140–143
   thermal modeling **1**: 144–149
Fabrication process. See *Molding; Manufacturing*
   process; and specific processes, e.g.,
   Injection molding.
Failure. See *Fracture.*
Failure analysis, introduction **4**: 3–6
Failure, competing micromechanisms . . **4**: 23, 35–36

Failure data

    carbon/epoxy composite **4**:76–77

    collection **4**:43–46

    cumulative probability diagram **4**:46–49

    statistical analysis **4**:46

    superposition **4**:108–110

Failure, environmental **4**:55–65

    ply, orthotropic laminate **5**:116–121

    shear, straight beam **5**:9

Failure mode, temperature effect **6**:14–15

Failure model, Tsai-Wu **5**:106–108

Failure prediction, unidirectional

    laminate **5**:114–121

Failure stress, manufacturing process . . .**4**:74–77, 80

Failure surface morphology

    carbon composite **4**:119–128

    glass fiber composite **4**:128–135

    PET **4**:130–135

    SMC **4**:135

Fastener. See *Joint(s), mechanically fastened.*

Fatigue, discontinuous fiber composite

    change in mechanical properties **1**:107–108

    failure process **1**:101–103

    fiber end effect **1**:100–101

    S-N relations **1**:103–107

Fatigue, life prediction **1**:107, 109–122

    data analysis **1**:114–120

    Weibull distribution **1**:122

Fatigue, multidirectional laminate

    boron/epoxy **1**:85–92

    compression effect **1**:92

    environmental effect **1**:98

    failure process **1**:85–90

    fatigue notch sensitivity **1**:96–97

    frequency effect **1**:94, 96

    graphite/epoxy **1**:88–92

    S-N relations **1**:90–92

    stacking sequence effect **1**:97–98

    temperature increase **1**:93–94

Fatigue, unidirectional laminate

    boron/epoxy **1**:77–84

    glass/epoxy **1**:77–78

    Kevlar/epoxy **1**:78–82

    longitudinal stress rupture **1**:79–82

    longitudinal tension fatigue **1**:75–79

    off-axis fatigue **1**:82–85

Fatigue of composites

    introduction **1**:73–75

    nomenclature **1**:73–74

Fatigue Wear. See *Wear mechanisms.*

Fiber

    aramid **1**:8–9

    asbestos **1**:4

    debonding **4**:25–31, 41–52

    E-glass **1**:5–6

    economic selection factors **3**:97

    fracture, stress and energy **4**:24–25

    general characteristics **1**:4–9, 17–18

    graphite **1**:6–8

    Kevlar **1**:8–9

    organic **1**:8

    S-glass **1**:6

    tensile strength **1**:4

Fiber bundle, failure model **4**:74–82

Fiber orientation, wear rate **4**:170–171

Fiber pull-out **4**:43–52

    shear stress **4**:28–29

Fiber reinforced plastics. See also

    Composite material(s).

    process comparison **3**:99–100

    properties, glass filled TP **3**:159–161, 166–169

    properties, graphite filled TP **3**:166–169

Fiber reinforcement, availability-cost

    index **3**:97, 100–101

Fiber volume fraction

    defined **1**:18–19

    test method **6**:7–8

Fiber weight fraction **1**:18–19

Fick's Law, transport property **2**:213

Filament, tensile test **6**:5–6

Filament winding

    equipment **3**:196–199

    history **3**:193

    misc. structures **3**:202–203

    pipe and tubing **3**:200–201

    tanks **3**:201–202

    winding patterns **3**:193–196

Fillers

    composition **1**:12–13

    for SMC **3**:133–134

    for thermosetting resins **3**:122–124

    general characteristics **1**:11–13

    properties **1**:12–13

Finite element analysis

    auto bumper **5**:238–243

    door intrusion beam **5**:230–232

First mode mechanism, stress transfer **5**:78–80

Flammability
polyester **3**:25–26
polyimides **3**:78
polysulfone **3**:49–50
Flexural failure
carbon-fiber composite **4**:122–124
Flexure test
comparison with tensile strength **6**:25–27
composite **6**:21–27
interlaminar fracture **6**:127–146
Fourier transform
bending **5**:162–166, 202–203
boundary problem solutions **5**:202–204
buckling **5**:181–187, 202–203
free vibration **5**:197–198
NDE digital system **6**:81–91
Fractography
definition and use **4**:119
Fracture
behavior, damage model **4**:101–110
energy **4**:87–88
fiber, stress and energy **4**:24–25, 35–36
flaw initiation **4**:69
graphite/epoxy laminate **4**:101–108
micromechanics **4**:3
microscopic models **4**:25, 41–43
single, ductile matrix **4**:17–20
single, short fiber **4**:15–17
single, stress-strain curve **4**:12–14
statistical aspects **4**:69–82
work of **4**:41–44, 50–52, 59–62
Fracture mechanics, interlaminar
energy release rates **6**:115–119
introduction **6**:113–119
mode I **6**:114–115, 119–126
mode II **6**:114–115, 126–146
stress intensity factors **6**:113–119
Fracture mechanics, macroscopic
composites application **4**:88–90
crack propagation **4**:85–97
fundamentals **4**:85–97
important equations **4**:86
test specimens **4**:90–94
unidirectional composites **4**:97–100
unidirectional materials **4**:97, 100
Fracture stress, specimen size effect **4**:73–74
Fracture toughness
carbon-fiber composites **4**:97–100

common unit conversions **4**:110
orthotropic materials **4**:95–97
Free edge. See *Edge effect.*
Free vibration. See also *Vibration.*
Fourier transform analysis **5**:197–198
Friction
adhesive component **4**:143–145, 160
coefficients **4**:165–175
plowing component **4**:145–146, 160
temperature influence **4**:146–148
test procedures **4**:159–164
Friction and wear. See also *Tribology.*
bearings **4**:165–167
behavior, general conclusions **4**:202–203
carbon fibers **4**:167, 170
fundamental aspects **4**:142–165
glass fibers **4**:167–168, 170
hybrid systems **4**:171–173
internal lubricants **4**:169
nomenclature **4**:139–140, 148–152

G

Galerkin method
bending problem **5**:163–166, 201–202
buckling problem **5**:178–181, 185–187, 201–202
free vibration analysis **5**:195–198, 200–201
Gantrez-treated **4**:55–56
Gating methods, NDE **6**:62–67, 79–81, 106
Gel coat, contact molding **3**:114
Glass/polyester
auto bumper beam **5**:234–243
tensile strength **3**:108
Glass/polyimide properties **3**:124
Glass fiber(s)
BMC **3**:10
contact molding **3**:106–108
coupling agent **3**:8
forms **3**:8–11
friction and wear **4**:167–168, 170, 183–191
manufacture **3**:7–8
preparation for molding **3**:8–11
prepreg **3**:10
size or finish **3**:7–8
SMC **3**:10–11
Glass fiber composite
door intrusion beam **5**:227–228
failure surface morphology **4**:128–135
friction and wear **4**:166–200

Glass-fiber/thermoplastics
  mechanical properties **3**:167
  physical properties **3**:166
Glass sphere
  wear behavior, reinforced TP **4**:183–191
Glass transition temperature **2**:149; **6**:11–12, 15
  moisture effect **6**:11–12
  plasticization effect **6**:11–12
Graphite fiber. See also *Carbon; Fiber.*
  contact molding **3**:107, 109
  manufacture **3**:8
  precursor **1**:7
  properties **1**:8
Graphite/bismaleimide **3**:69–72
Graphite/epoxy composite
  environmental failure **4**:55–60
  fatigue **1**:84–100
  flexure strength **6**:14
  fracture behavior **4**:101–108
  friction and wear **4**:166–200
  LEFM **4**:97–100
  moisture effect **6**:9–11, 13–14
  representative properties **1**:59, 68
  thermal conductivity **4**:147
  transverse strength, angle bracket **5**:41–42
  truck applications **5**:207–226
  unidirectional properties **1**:149
Graphite/PEEK **3**:56–60
Graphite/polyimide **3**:60, 74–78
Graphite/polysulfone **3**:53–56
Graphite/thermoplastic **3**:166–169

H

Halpin-Tsai equation **2**:56, 82–86, 107–108, 220
Hand lay-up molding. See also *Contact molding.*
  curing **3**:112–114
  gel coat **3**:114
  styrene emission **3**:105, 115
  trimming **3**:114
Heat deflection temperature (HDT)
  engineering thermoplastics **3**:39–40, 45
  epoxy resins **3**:35–39
  polyester resin **3**:25
Heat distortion temperature **6**:12
Hoffman criterion, interaction
    coefficient **5**:108–109

Hooke's law
  anisotropic material **2**:4–11
Hybrid system, friction and wear **4**:171–173

I

IITRI compression test **6**:29, 31–32
Impact failure
  carbon-fiber composite **4**:125
Impact resistance **1**:4
Injection molding, thermoplastics
  fiber content effect **3**:159–161
  FRTP properties **3**:159–161, 166–169
  machine design and control **3**:152–155
  machine operation **3**:155–158, 163–165
  mold design **3**:158, 163–165
  molding filled materials **3**:158–159
  molding machine principles **3**:149–158
  plasticization **3**:155–157
  precision molding **3**:165–166
  shrinkage **3**:165–166
Injection molding, thermosets
  bulk molding compound **3**:176–178
  general principles **3**:95, 173
  processing **3**:174–175
  tool design **3**:175
Injection-compression molding, thermoset **3**:177
In-plane engineering constants. See *Engineering*
    constants, in-plane.
Inspection, nondestructive, ultrasonic **6**:57–109
Interaction coefficient
  Hoffman criterion **5**:108–109
  Tsai-Hill criterion **5**:108
Interlaminar fracture
  mode I **6**:114–115, 119–126
  mode II **6**:114–115, 126–146
  mode III **6**:114–115, 146–158
Interlaminar shear, mode III **6**:146–158
Interlaminar stress **5**:71
  edge effect **5**:77–93
  stacking effect **5**:88–93
Introduction to
  composite materials **3**:3
  compression molding **3**:93, 119
  contact molding **3**:89–90, 105
  failure analysis **4**:3–6
  fatigue **1**:73–75

filament winding **3**:193
interlaminar fracture mechanics **6**:113-119
mechanics of composite materials **1**:17
pultrusion **3**:207
rheology **3**:217
sheet molding compound **3**:133
thermoplastic matrix resins **3**:39-40
transport properties **2**:205-206
Isotropy, composite materials **2**:3, 24-26

J

Joint. See also *Attachment.*
Joint(s)
adhesively bonded **5**:27-32
buffer strip **5**:26
generic types **5**:22-23
mechanically fastened **5**:23-27, 31-42
selection criteria **5**:21-23

K

***k* Factor, shear deformation** **5**:155-156
Kevlar aramid fiber. See also *Fiber.*
contact molding **3**:107, 109
general characteristics **3**:8
Kinetics, moisture absorption **4**:63
Knit line **1**:3

L

Laminate. See also *Composite material(s);*
Unidirectional composite(s).
bending-stretching coupling **5**:62-63
constitutive relations **5**:60-62
elastic constants **5**:65-66
moisture effect on properties **6**:11-19
nomenclature **5**:59
quasi-isotropic ply stress **5**:67-71
quasi-isotropy **5**:66-67
symmetric, bending **5**:162-166
symmetric, buckling **5**:178-187
symmetric, free vibration **5**:196-198
temperature effect on properties **6**:14-19
transport properties **5**:74-75

unsymmetric, bending **5**:166-171
unsymmetric, buckling **5**:187-193
unsymmetric, free vibration **5**:198-201
Laminate strain, moisture. See *Expansional strains.*
Laminate strain, temperature.
See *Expansional strains.*
Lamination theory
analogy, short fiber composite **5**:75-77
basic assumptions **5**:59-60
short fiber composite **5**:75-77
Laplace transform **2**:190-200
Leaf spring, heavy truck **5**:219-226
LEFM. See *Linear elastic fracture mechanics.*
Linear elastic fracture mechanics
shear lag analysis **4**:99-101
unidirectional composites **4**:97, 100
Loading, strength prediction **5**:105-121
Lockheed compression test **6**:32
Lubricants
internal, friction and wear **4**:169
thermosetting resins **3**:123

M

Manufacturing process. See also *Molding;*
Processing.
failure stresses **4**:74-77, 80
polyester resin **3**:15-16
sheet molding compound **2**:95-97
MARC computer code analysis **5**:51-54
Matched die molding **3**:91
Material behavior (anisotropic)
directional dependence **2**:3, 10-18
hexagonal symmetry **2**:24
isotropy **2**:24-26
orthotropic symmetry **2**:20-22
rotational transformation **2**:12-14
square symmetry **2**:22-24
Material characterization, ASTM tests **6**:4
Material properties. See also *Composite*
materials; Properties, composite; Properties,
resin; and specific materials, e.g.,
Graphite/epoxy.
definitions **1**:49-55
design allowables **1**:59
size effect **1**:60-62
variability **1**:55-59

**Materials, homogeneous, strength
        analysis** **5**: 105–114
Matrix
    cracking **4**: 33–34
    plastic deformation **4**: 31–33
    tensile test **6**: 6–7
**Maximum strain criterion, strength
        analysis** **5**: 106
**Maximum stress criterion, stress
        analysis** **5**: 105–106
Metalloplastic
    definition **2**: 206
    electrical property model **2**: 206
Methodology, composite design **5**: 3–5
Microstructure, C-scan **6**: 102–109
MIL Hdbk 5 . . . .**1**: 59
Mode I
    interlaminar fracture **6**: 114–115, 119–126
Mode II
    interlaminar fracture **6**: 114–115, 126–146
Mode III
    interlaminar shear **6**: 114–115, 146–158
Model, fatigue life prediction **1**: 107, 109–122
Model, fiber bundle failure **4**: 74–77
Model, mechanical behavior, particular composites
    bounding **2**: 101–102
    embedding **2**: 100, 103–107
    mechanics of materials **2**: 99–100
    Reuss **2**: 101–102
    semi-empirical model **2**: 107–108
    Voigt **2**: 101–102
**Model, mechanical behavior, sheet
        molding compound**
    aggregate **2**: 119–123
    micro-laminate **2**: 123–135
    Reuss **2**: 119–123
    Voigt **2**: 120–121
Model, microscopic fracture **4**: 23–25,
        35–36, 41–43
Model, unidirectional materials behavior
    bounding model **2**: 73–82, 86–89
    Halpin-Tsai equation **2**: 56, 82–89
    mechanics of materials **2**: 56–65, 86–89
    rule of mixtures **2**: 53–56, 65
    self-consistent field **2**: 65–73, 86
Model, viscoelastic behavior, mechanical analogs
    Boltzmann superposition principle **2**: 171–173
    differential formulation **2**: 167–171
    failure, Tsai-Wu **5**: 106–108

    integral formulation **2**: 171–179
    linking relationships **2**: 179–186
    Maxwell **2**: 159–166
    three-parameter viscoelastic behavior . .**2**: 163–166
    two-parameter viscoelastic behavior . . .**2**: 159–163
    viscoelastic behavior, mathematical **2**: 166–181
    viscoelastic descriptors **2**: 155–157, 172
    viscoelasticity, linear, constitutive
        relationships **2**: 182–186
    Voight-Kelvin model **2**: 159–166
**Model, viscoelastic behavior, transform
        approach** **2**: 190–200
Model, wear process **4**: 152–159
**Models, strength analysis, homogeneous
        materials** **5**: 105–114
**Models, thermomechanical behavior,
        fabric composites**
    bridging model **1**: 139–140, 143–149
    crimp model **1**: 136–139, 143–149
    mosaic model **1**: 133–136, 143–149
Model transport properties
    comparison with characterization
        data **2**: 225–230
    formulation **2**: 222–224
    particulate systems **2**: 224–225
Moisture, laminate strain. See *Expansional strains.*
Moisture absorption
    coefficient of expansion **6**: 9–10
    effect on properties **6**: 11–19
    effect on *Tg* **6**: 11–12
    epoxy resin **3**: 39
    kinetics **4**: 63
    measurement **6**: 12–14
Mold. See *Tooling.*
Mold design
    compression **3**: 126–127, 141
Molding. See *Autoclave molding; Compression
        molding; Contact molding; Hand lay-up
        molding; Injection molding; Matched die
        molding; Open molding; Press molding;
        Processing; Resin injection molding;
        Resin transfer molding; Spray-up molding;
        Transfer molding.*
Molding equipment
    compression molding **3**: 127, 140–141
    filament winding **3**: 196–199
    injection molding **3**: 149–158
    pultrusion **3**: 207–211
    SMC **3**: 140–141

Molding processes
economic factors....................**3**:99–101
overview.........................**3**:89–100
SMC, DMC, BMC.............**3**:93, 139–143
Morphology
failure surface, carbon composite......**4**:119–128
failure surface, glass-fiber composite . . **4**:128–135
PET.............................**4**:130–135
Multi-directional laminate
failure process....................**1**:85–90
fatigue............................**1**:85–98
S-N relations......................**1**:90–92

N

NASTRAN computer code analysis.......**5**:54–55
NDE, NDI, NDT. See *Nondestructive evaluation.*
Netting analysis, filament winding......**3**:194–195
NIST compression test....................**6**:29
Nomex...................................**1**:8
Nondestructive evaluation, ultrasonic....**6**:57–109
analog system.........................**6**:57
delamination.......................**6**:72–75
digital system....................**6**:58, 79–97
A-scan...........................**6**:58–61
B-scan...............................**6**:62
C-scan.............**6**:58, 62–75, 78, 93–109
Northrup compression test.................**6**:29

O

Open molding. See *Contact molding.*
Orthotropic laminate. See also *Specially orthotropic plate.*
ply failure.........................**5**:116–121
Overview, composite design methodology ... **5**:3–5
Oxygen index
poly(amide-imide).....................**3**:50
polyphenylene sulfide..................**3**:50

P

PAN. See *Polyacrylonitrile.*
Paris relation...........................**4**:95
Particulate filled composite
bounding relationships..............**2**:101–102
combining rules....................**2**:108–114
embedding model..................**2**:103–107
Lewis-Nielsen equation.............**2**:110, 116

modeling, mechanical behavior........**2**:99–116
Poisson's ratio..........................**2**:93
prediction of properties..............**2**:99–116
semi-empirical models..............**2**:107–108
Young's modulus.............**2**:93, 99–107, 110
PET. See *Polyethylene terephthalate.*
Phenolic resins
B- and C-stage........................**3**:121
commercial availability..................**3**:84
composition.....................**3**:79–82, 119
cure.......................**3**:82–83, 119–121
flammability and smoke...................**3**:83
prepreg................................**3**:84
properties, cured resin...................**3**:83
properties, laminate....................**3**:82
Pipe and tubing
filament winding...................**3**:200–201
Plastic deformation.....................**4**:31–33
Plasticization, T_g **effect**.....................**6**:11–12
Plate
angle ply, bending stiffness............**5**:93–94
angle ply, coupling stiffness...........**5**:94–96
anisotropic, bending and buckling.....**5**:125–129
anisotropic, displacement formation ...**5**:127–129
anisotropic, one-dimensional theory ...**5**:139–143
anisotropic, stress function
formulation....................**5**:128–129
bending, transverse loading.........**5**:159–171
buckling and vibration.............**5**:142–143
cylindrical bending................**5**:139–142
energy principle application.........**5**:129–136
specially orthotropic..............**5**:159–162,
171–178, 194–196
stiffnesses, transformation relations.....**5**:64–65
thin, stress calculation...............**5**:18–21
Ply
failure, orthotropic laminate..........**5**:116–121
stress, quasi-isotropic laminate........**5**:67–71
Poisson's ratio
in-plane..............................**2**:59
micromechanical model...........**2**:52, 59, 67
unidirectional composite................**1**:20
Polyacrylonitrile, graphite fiber precursor.....**1**:7
Poly(amide-imide)
electrical properties....................**3**:51
oxygen index.........................**3**:50
processing............................**3**:51
properties, cured resin...............**3**:45–46
structure.............................**3**:41

Polyblend, combining rule **2**: 115
Polycarbonate
   electrical properties **3**: 51
   processing **3**: 51
   properties, composite **3**: 167–168
   properties, cured resin **3**: 45
   structure **3**: 41
Polyester resin(s)
   backbone structure effect **3**: 17–18
   bisphenol A/fumarate **3**: 22–23
   chemical resistance **3**: 24
   commercial availability **3**: 26–27
   composition or structure **3**: 15–16
   contact molding **3**: 109–111
   cured properties **3**: 23–25
   curing **3**: 19–22
   flammability and smoke **3**: 25–26
   manufacture **3**: 15–16
   monomer function **3**: 18–19
   thermal properties **3**: 25
   vinyl ester resin **3**: 22–23
Polyetherether ketone (PEEK)
   chemical resistance **3**: 48
   processing **3**: 51–53
   properties, composite **3**: 44, 56–60
   properties, cured resin **3**: 45–46
   structure **3**: 41–44
Polyetherimide
   processing **3**: 51
   properties, neat resin **3**: 45–46
   structure **3**: 41
Polyethylene terephthalate (PET)
   failure surface morphology **4**: 130–135
Polyimides
   acetylene capped **3**: 67
   addition type chemistry **3**: 62, 64
   commercial availability **3**: 78
   condensation type chemistry **3**: 61–63
   electrical properties **3**: 76–78
   environmental resistance **3**: 75–77
   flammability and smoke **3**: 78
   graphite composite properties **3**: 74–78
   in compression molding **3**: 122
   LARC-160 .... **3**: 65–66, 71
   PMR-II **3**: 67, 71
   PMR-15 .... **3**: 65–66, 71
   processing **3**: 65–69
   properties, laminate **3**: 69–74
   properties, neat resin **3**: 69–70, 72
   thermoplastic, melt processable **3**: 67–69

Polyphenylene sulfide
   chemical resistance **3**: 47
   electrical properties **3**: 51
   oxygen index **3**: 50
   processing **3**: 51–52
   properties, composite **3**: 167–168
   properties, cured resin **3**: 45
   structure **3**: 41, 44
Polysulfone
   adhesive bonding **3**: 60–61
   chemical resistance **3**: 47
   electrical properties **3**: 50–51
   processing **3**: 50–52
   properties, composite ... **3**: 40, 47, 55–56, 167–168
   properties, neat resin **3**: 45, 48
   smoke and flammability **3**: 49–50
   structure **3**: 41
   vs epoxies **3**: 42, 53–55
Polyurethane, RRIM **3**: 181, 187–189
Postcure, thermosetting resins **3**: 128
Postmolding. See *Postcure.*
Precision molding
   reinforced thermoplastics **3**: 165–166
Precursor, graphite fiber **1**: 7
Prediction
   fatigue life **1**: 107–122
   strength, combined loading **5**: 105–121
Prediction, mechanical properties
   bounding technique **2**: 190–200
   inversion of bounds **2**: 197–199
   viscoelastic bounds **2**: 195–197
Prediction of properties
   particulate filled composite **2**: 99–116
   sheet molding compound **2**: 118–137
   unidirectional composite **2**: 52–56, 71–83, 86
Prepreg
   available forms **3**: 10
   epoxy **3**: 27
   PEEK **3**: 57
   phenolic **3**: 84
   polyimide **3**: 62, 65, 67–69, 78
   thermoplastic matrix **3**: 42, 53
Press molding **3**: 95
Processing. See also *Curing; Molding.*
   BMC **3**: 93, 139–143, 145, 176–177
   fiber reinforced plastics, comparison ... **3**: 99–100
   filament winding **3**: 193–196
   injection molded thermoplastics **3**: 155–159, 163–165
   injection molded thermosets **3**: 174–175

poly(amide-imide) **3**:51
polycarbonate **3**:51
polyetherether ketone **3**:51–53
polyetherimide **3**:51
polyimides **3**:65–69
polyphenylene **3**:51–52
polysulfone **3**:50–52
RRIM **3**:182–185
SMC **3**:93, 139–143

Properties, composite
abrasiveness **4**:174–179
alumina/epoxy **1**:69
aramid/epoxy **1**:67
BMC **3**:143–144
boron/epoxy **1**:67
E glass/epoxy **1**:66
fiber reinforced thermoplastic . **3**:159–161, 166–169
given material system **6**:3
glass filled thermoplastic **3**:159–161, 166–169
graphite/epoxy **1**:59, 68
graphite filled thermoplastic **3**:166–169
laminate, moisture effect **6**:11–19
laminate, temperature effect **6**:14–19
phenolic **3**:82
polycarbonate **3**:167–168
polyetherether ketone **3**:44, 56–60
polyimide **3**:69–74
polyphenylene **3**:167–168
polysulfone **3**:40, 47, 55, 167
RRIM urethane **3**:187–189
S glass **1**:66
SMC **3**:143–144
unidirectional materials **1**:62–69, 149
wear **4**:187–191

Properties, resin
epoxy, cured **3**:35–39
phenolic, cured **3**:83
poly(amide-imide), neat **3**:45–46
polycarbonate, neat **3**:45
polyester, cured **3**:24–25
polyetherether ketone, neat **3**:45–46
polyetherimide, neat **3**:45–46
polyimide, neat **3**:69–70, 72
polyphenylene, neat **3**:45
polysulfone, neat **3**:45, 48

Pultrusion
curing **3**:211
equipment **3**:207–211
introduction **3**:207
overview **3**:96–97

resin impregnation **3**:209–210
structures design **3**:212–213

Q

**Quadratic interaction criterion, strength
analysis** **5**:106–114
Quality control
SMC **3**:138–139, 143
Quasi-isotropic laminate
bending stress **5**:68–71
defined **1**:37
ply stress **5**:66–71

R

R-curve effect, interlaminar fracture . . . **6**:125–126
Rail, truck frame, design study **5**:207–214
Rail shear test. See *Crack rail shear test.*
Regrind material, transfer molding **3**:129
Reinforced Reaction Injection Molding (RRIM)
choice of basic materials **3**:181–182
general description **3**:181
mechanical properties **3**:187–189
metering and mixing **3**:184
mold design **3**:185–187
processing **3**:183–185
Reinforcements. See also *Fiber; Roving.*
for contact molding **3**:106–109
for filament winding **3**:193–195
for RRIM **3**:183–184
for SMC **3**:135
for thermoplastic resins **3**:159–161
for thermosetting resins **3**:123–124, 177–178
forms **3**:8–11
general fiber characteristics **1**:4
glass fibers **1**:5–6; **3**:7–8
graphite (carbon) fibers **1**:6–8; **3**:8
Kevlar **1**:8–9; **3**:8, 107–108
mechanics **1**:17–44, 131–149
Residual stress **5**:73–74
Resin(s), general characteristics **1**:9–11
thermoplastics **1**:10–11
thermosets **1**:10
Resin injection molding **3**:91–92
Resin transfer molding. See *Resin injection
molding.*

Resins, availability-cost index **3:** 97–98, 100–101
Resonant frequencies, composite beam **5:** 14–16
Reuss model, particulate system **2:** 101–102
Rheology, fiber filled polymers
    inhomogeneous suspensions **3:** 235–237
    introduction **3:** 217
    long fiber suspensions **3:** 227–231
    low viscosity suspensions **3:** 231–234
    monodispersed suspensions **3:** 217–220
    particle distribution effects **3:** 220–221
    particle geometry effects **3:** 222–227
    spherical particle suspensions **3:** 217–227
Roving. See also *Woven roving* **3:** 9, 107, 114, 123, 135, 159–162
RRIM. See *Reinforced Reaction Injection Molding.*

S

S-basis, statistical allowable **6:** 4–5
S-glass/epoxy
    representative properties **1:** 66
S-glass fiber **1:** 6
    economic factors **3:** 97
S-N relations
    composite constituents **1:** 76
    discontinuous fiber laminate **1:** 103–107
    frequency effect **1:** 94–96
    metal matrix composites **1:** 76–79
    multi-directional laminate **1:** 90–92
    unidirectional laminate **1:** 77–78
Safety
    materials handling, RRIM **3:** 185–187
Sandwich, compression test **6:** 32–35
Second mode mechanism, stress transfer .. **5:** 80–88
Selection criteria
    joints **5:** 21–23
Shear
    coupling, anisotropic material **2:** 42–43
    deformation, *k* factor **5:** 155–156
    failure, straight beam **5:** 9
    modulus, in-plane **2:** 60–61
    short beam specimen **5:** 148–149
    stress, fiber pull-out **4:** 28–29
    transverse deformation **5:** 149–154
Shear lag analysis, LEFM **4:** 99–100
Shear strength
    carbon fiber composite **4:** 127–128
    unidirectional composite **1:** 28–31

Shear test, composite
    biaxial loading **6:** 48–52
    coupon test **6:** 38–40
    interlaminar, mode III **6:** 114–115, 146–158
    off-axis tensile test **6:** 40–42
    rail shear test **6:** 42–44
    short beam test **6:** 47–48
    torsion test **6:** 44–47, 127
Sheet molding compound (SMC)
    aggregate models **2:** 119–123
    composition **3:** 10–11, 133–134
    compounding **3:** 134–136
    design **2:** 137–141
    failure surface morphology **4:** 135
    introduction **3:** 133
    manufacturing process **2:** 137–141
    maturation **3:** 136
    mechanical properties **3:** 143–144
    micro-laminate model **2:** 93–98, 123–125
    mixing **3:** 135–136
    modeling, mechanical behavior **2:** 118–137
    molding equipment **3:** 140–141
    molding technology **3:** 93, 139–143
    Poisson's ratio **2:** 133, 136–140
    prediction of properties **2:** 118–137
    processing influence on properties **3:** 145
    quality control **3:** 138–139, 143
    reinforcement variations **3:** 136–138
    Young's modulus **2:** 133, 136–140
Short beam shear specimen **5:** 148–149
Short fiber composite
    lamination theory, analog **5:** 75–77
SMC. See *Sheet molding compound (SMC).*
Smoke
    phenolics **3:** 83
    polyester **3:** 25–26
    polyimides **3:** 78
    polysulfone **3:** 49–50
Specially orthotropic plate **5:** 194–196
    bending **5:** 159–162
    buckling **5:** 171–178
Specimen size effect
    fracture stress **4:** 73–74
Spray-up molding **3:** 90–91, 114–115
Stacking effect, interlaminar stress **5:** 88–93
Statistical allowable
    A-basis **6:** 4–5
    B-basis **6:** 4–5
    S-basis **6:** 4–5
    typical basis **6:** 4–5

Statistical analysis, extreme value **4**:59-62
Stiffness
    definition **4**:3
    degradation, temperature aging **4**:55-58
Stiffness, bending. See *Bending stiffness.*
Stiffness, coupling. See *Coupling stiffness.*
Strength, prediction, combined
      loading **5**:105-121
Strength analysis
    maximum strain criterion **5**:106
    maximum stress criterion **5**:105-106
    quadratic interaction criterion **5**:106-114
Strength degradation
    failure model **1**:112-117
Stress
    bending, quasi-isotropic laminate **5**:68-71
    curved beam **5**:12-14
    edge **5**:98-101
    intensity factors **6**:113-119
    interlaminar **5**:71
    interlaminar, edge effect **5**:77-93
    plane **2**:40-42
    residual **5**:73-74
    straight beam **5**:9-11
    thin plate **5**:18-21
    transfer, first mode mechanism **5**:78-80
    transfer, second mode mechanism **5**:80-88
Stress-corrosion effect **4**:65
Stress-life relations. See *S-N relations; Fatigue.*
Stress-strain behavior
    woven fabric **1**:140-143
Styrene emission
    contact molding **3**:105, 115
Superposition, failure data **4**:108-110
SWRI compression test **6**:32
Symmetry, constitutive relation **5**:63

T

Tanks
    filament winding **3**:201-202
Temperature, laminate strain.
    See *Expansional strains.*
Temperature effect
    on failure mode **6**:14-15
    on properties **6**:14-19
Tensile failure
    carbon-fiber composite **4**:119-122

Tensile-flexure strength comparison **6**:25-27
Tensile strength
    unidirectional composite **1**:24-27
Tensile test
    composite **6**:19-21
    composite, off-axis **6**:27-29
    matrix **6**:6-7
    single filament **6**:5-6
Test. See specific property, e.g., *Shear test.*
Test methods, composite material
    ASTM **6**:4
    compression **6**:29-35
    crack rail shear **6**:147-158
    density **6**:7
    end notch flexure **6**:127-146
    shear **6**:35-48
    short beam shear **6**:127
    tensile **6**:19-21
    volume fraction **6**:7-8
Test procedures
    friction **4**:159-164
    wear **4**:159-164
Test specimen
    fracture mechanics **4**:90-94
Testing, moisture effect **6**:11-19
T_g. See *Glass transition temperature.*
Thermal expansion coefficient **6**:8-9
Thermoplastic resins. See also specific materials,
      e.g., *Polyetherether ketone; Polysulfone.*
    adhesive bonding **3**:60-61
    advantages vs thermosets **3**:40-43
    definition **1**:10
    engineering thermoplastics **3**:43-50
    high temperature **3**:43-50
    introduction **3**:39-40
    performance vs epoxies **3**:53-60
    processing **3**:50-53
    typical properties **1**:10-11
    wear behavior, glass fiber reinforced ... **4**:183-191
    wear behavior, glass sphere
      reinforced **4**:183-191
Thermoset, fiber reinforced, abrasive wear
Thermoset injection-compression molding ... **3**:177
Thermoset injection molding **3**:95
    behavior **4**:191-201
Thermosetting resins. See also specific materials,
      e.g., *Epoxy; Polyester; Polyimides.*
    definition **1**:10
    typical properties **1**:10-11
Thick molding compound (TMC) **3**:137

TMC. See *Thick molding compound.*

Tooling

compression molding **3:**126–127

contact molding **3:**105–106

injection molded thermoplastic . . . **3:**158, 163–165

injection molded thermoset **3:**175

Pultrusion **3:**211

RRIM **3:**185–187

SMC and BMC **3:**141

transfer molding **3:**129

Torsional frequency, beam **5:**16–18

Torsional loading, drive shaft **5:**215–218

Toughness. See *Work of fracture.*

Trade-off study, composite design **5:**3

Transducer, ultrasonic **6:**75–78

Transfer molding **3:**93, 128–129, 177

Transport behavior, 3-dimensional **2:**215–218

Transport properties

combining rules **2:**218–222

directional dependence **2:**216–218

electrical properties **2:**207–214

elementary material descriptors **2:**206–215

introduction **2:**205–206

laminate **5:**74–75

magnetic permeability **2:**211–213

models **2:**222–225

percolation process **2:**205–206, 221–222

thermal conductivity **2:**212–213

unidirectional composite **2:**69–70

Transport relationships, comparison **2:**214

Transverse loading

plate bending **5:**159–171

Transverse properties

unidirectional composites **1:**28–31

Tribology. See also *Friction and wear.*

behavior, elevated temperatures **4:**172–175

terms and definitions **4:**141, 148–152

Tsai-Hill criterion, interaction coefficient **5:**108

Tsai-Wu failure model **5:**106–108

Tubing, See *Pipe* **3:**200–201

Typical basis, statistical allowable **6:**4–5

U

Ultrasonic inspection

analog system **6:**57

digital system **6:**58, 79–97

A-scan **6:**58–61

B-scan **6:**62

C-scan **6:**62–78, 93–109

Unidirectional composite(s). See also

Composite materials.

axial strength properties **1:**23–28

bounding methods, material behavior **2:**73–82

elastic properties **1:**19–23

elastic properties, mechanics analysis . . . **2:**56–65

engineering constants **2:**56–72

failure prediction **5:**114–121

fatigue **1:**75–85

fracture mechanics **4:**97, 100

prediction of properties **2:**52–56, 71–83, 86

properties, through-the-thickness **1:**52–53

stress-strain properties **1:**62–65

transport properties **2:**69–71

transverse strength properties **1:**28–31

typical, epoxy matrix systems **1:**65–69

Young's modulus **1:**20–22

V

Vibration

anisotropic plate **5:**125–129, 142–143

Fourier series analysis **5:**197–198

Galerkin method **5:**195–198, 200–201

laminated plate analysis **5:**193–201

symmetric laminate **5:**196–198

unsymmetric laminate **5:**198–201

Vinyl ester resin **1:**10–11; **3:**22–23

contact molding **3:**112

Viscoelastic behavior

dynamic response **2:**150–157

instantaneous (static)

response **2:**146–149, 155–157

introduction **2:**145–146

prediction, model **2:**189–200

time-temperature shift factor **2:**154–155

transient response **2:**149, 155–157

Viscoelastic minimum energy principle **2:**190

Viscoelastic solid, linear, definition **2:**191–192

Voigt model, particulate system **2:**101–102

Volume fraction

fiber **1:**18–19; **6:**7–8

SMC **2:**133–135

Von Mises yield criterion, plane stress . . .**5:**109–114

W

Wear
    abrasive, general conclusions **4**:203
    behavior, GF-PETP and
        GS/GF/CF-PA **4**:183–190
    fiber reinforced thermoplastic **4**:183–191
    fiber reinforced thermoset **4**:191–201
    glass sphere reinforced TP **4**:183–191
    material property data **4**:187–191
    mechanisms **4**:148–152
    properties **4**:165–200
    quantitative mechanisms **4**:148–152
    testing **4**:159–164
Wear and friction. See also *Tribology.*
    bearings **4**:165–167
    behavior, general conclusions **4**:202–203
    carbon fiber composite **4**:167, 170
    fundamental aspects **4**:142–165
    glass fiber composite **4**:167–168, 170
    hybrid systems **4**:171–173
    internal lubricants **4**:169
    nomenclature **4**:139–140, 148–152
Wear mechanisms **4**:148–152
    abrasive mechanism **4**:175–179
    microscopy **4**:183–189, 192–193, 197–198

Wear process, model **4**:152–159
Wear rate, fiber orientation **4**:170–171
Weave, fabric **1**:32–33
Weibull distribution
    failure statistics **4**:70–82
    fatigue life data **1**:122
Weight fraction
    fiber **1**:18–19
    SMC **2**:133–135
Westergaard stress function **4**:110–114
Williams-Lauder-Ferry equation **2**:154
Work of fracture **4**:41–44, 50–52, 59–62
Woven fabric. See *Fabric, woven.*
Woven roving **1**:32–33; **3**:107–109

Y

Yield criterion, Von Mises **5**:109–114
Young's modulus
    in-plane **2**:56–58, 60
    micromechanical model **2**:50, 56–58, 60
    particulate composite **2**:93, 99–107, 110
    sheet molding compound **2**:133, 136–140
    unidirectional composite **1**:20–22